Learn Algebra NOW!

Algebra for the Person Who Has Never Understood Math!

Minute Help Guides

Minute Help Press

www.minutehelp.com

© 2011. All Rights Reserved.

Table of Contents

INTRODUCTION 3

CHAPTER 1: ALGEBRAIC EXPRESSIONS 4

    TERMINOLOGY 5
    USING EXPRESSIONS 6
    ADDING ALGEBRAIC TERMS 6
    SUBTRACTING ALGEBRAIC TERMS 7
    MULTIPLYING ALGEBRAIC TERMS 7
    DIVIDING ALGEBRAIC TERMS 8
    MORE ON MULTIPLYING AND DIVIDING ALGEBRAIC TERMS 8
    ALGEBRAIC TERMS RAISED TO POWERS 9
    SQUARE ROOTS OF ALGEBRAIC TERMS 11
    CONCLUSION 11

CHAPTER 2: EQUATIONS AND INEQUALITIES 12

    EQUATIONS 12
    INEQUALITIES 15

CHAPTER 3: RELATIONS & FUNCTIONS 19

    RELATIONS 19
    FUNCTIONS 20
    FUNCTION RULES 23
    WRITING FUNCTION RULES 25

CHAPTER 4: POLYNOMIAL AND RATIONAL FUNCTIONS 28

    POLYNOMIAL FUNCTIONS 28
    STANDARD FORM 28
    NAMING POLYNOMIALS 29
    WORKING WITH POLYNOMIALS 29
    PUTTING IT ALL TOGETHER 30
    RATIONAL FUNCTIONS 31
    ADDING AND SUBTRACTING RATIONAL EXPRESSIONS 33

CHAPTER 5: SYSTEMS OF EQUATIONS 35

    GRAPHING 35
    SUBSTITUTION 38
    ELIMINATION 39
    SYSTEMS OF INEQUALITIES 42
    CONCLUSION 44

CHAPTER 6: EXPONENTIAL AND LOGARITHMIC FUNCTIONS 45

    EXPONENTIAL FUNCTIONS 45
    LOGARITHMIC FUNCTIONS 48

CONCLUSION 53

Introduction

Mathematics is a fascinating field. The ancient Greeks were so mesmerized by it that they formed a religion around the worship of numbers. Today, mathematics is used all around us in everything from baking to electrical engineering. There's no reason not to be totally in love with it, right?

Well, not exactly. Despite its usefulness, math can be very difficult. Many people find this to be both frustrating and discouraging, especially when they first encounter algebra.

This mathematics guide is aimed towards helping you understand—and hopefully enjoy—the fundamentals of algebra. The topics covered include algebraic expressions, equations and inequalities, relations and functions, polynomial and rational functions, exponential and rational functions, and systems of equations and inequalities. I know that sounds like a lot, but we're going to cover every topic at a gentle pace, showing how all of these topics are related and how they can be understood with just a little bit of work. (Sorry, but there's no way around doing work if you want to understand algebra.) By the time you finish reading this guide, you should love math as much the Greeks who worshipped it – well, you'll at least come out of it understanding the basics of algebra!

Chapter 1: Algebraic Expressions

Terminology

In this section we're going to discuss algebraic expressions. To do that, we'll need to start with a definition. Now, before I give you my definition of this term, let me just say this: don't worry. A lot of what we are going to do in this section is break down the definition and clearly explain what it means. This means that we're going to cover a lot of terminology. After covering some basic definitions, we'll move on to talk about some really useful rules to know about algebraic expressions. Okay, here it is: an algebraic expression is a phrase consisting of one or more algebraic terms.

This definition probably means nothing to you if you don't understand two key concepts: phrases and terms. Phrases are a lot easier to understand, so we'll start there. In math, a phrase is simply a collection of symbols that makes sense. Phrases in math are a lot like sentences in English. You could say, "two times a number" (2x), "one plus one equals two" (1 + 1 = 2), or "one third plus an unknown number equals nine" (1/3 + x = 9). As you can see, you could say the same phrases with math. Painless, right?

Algebraic terms are easy to understand as well, but talking about them is going to lead us to discuss other words you've probably heard thrown around the classroom but might not fully understand. An algebraic term is the product of a constant and one or more variables. A constant is a number and a variable is a letter that stands for an unknown number. So 2y, 4xy, and 4a are all algebraic terms.

All of the examples of algebraic terms given so far have been monomials, or expressions with just one algebraic term. These terms have parts—take 2y for example. 2 is a coefficient, or number that comes before the variable. y is the variable.

Now that we know what algebraic expressions are, it might be useful to learn how to talk about them. Algebraic expressions with more than one term are called polynomials. Examples of polynomials are 2y + 1, (xy + xz) + 1, and $x^2 + 2xy + y^2$. Polynomials with two terms are called binomials. Can you tell which of the polynomials listed above is a binomial? If you answered 2y + 1, give yourself a pat on the back.

Got it? Good. You should be an expert on the terminology associated with algebra by now. But just reading is never enough in math. Math, like most things, is learned best when it's put to use. So fill out this quick quiz and test your knowledge before you move on.

Quick Quiz:

1. In the algebraic term 2x, what is the coefficient and what is the variable?
2. A monomial has how many terms?
3. What would x + y be best described as?
4. How many terms are in the phrase $a + 2x + 5x^2$?
5. What is an algebraic expression with more than one term called?

Solutions:

1. 2 is the coefficient, x is the variable.
2. One
3. A binomial
4. Three
5. A polynomial

Using Expressions

By now we should know what algebraic expressions are. Now we get to answer the question, "Why on Earth do they exist and why do we use them?" Well, the answer is…simplicity. Algebraic expressions are a really useful way to make meaningful sentences in math when you have unknown quantities. We get algebraic expressions by applying the basic rules of arithmetic to unknown numbers. At its core, this is not a difficult thing to do. Actually, it can be as easy as $1 + 1 = 2$.

Don't believe me? Let's see:

Adding Algebraic Terms

$1 + 1 = 2$, right?

Well, what would we do if we didn't know one of those numbers? We would put:

$1 + x = 2$

Not bad, right? Now, say we didn't know either number. Say for example, I have two identical bolts and I don't know how much they weigh. I could say that both of them together weigh:

$x + x$

Here's an example of where we can apply the rules of arithmetic to numbers we don't know. What does $x + x$ equal?

$2x$.

See, $x + x$ is as easy as $1 + 1$.

Another example:

What if the number were different? What if my bolts weigh different amounts? In this case, we would say that together my bolts weigh:

$x + y$

We use different letters here because we're working with different amounts. And what does $x + y$ equal?

$x + y$

It's really that easy. The reason that we can do x + x and not x + y is that we know that two of any number is going to be the same as two times that number.

What we have just learned here is that with addition we can combine like terms or add up coefficients when they have the same terms. As we will see in a moment, this works with subtraction as well. Before we move on, however, let's revisit some old arithmetic. In our example, we added x and x and got 2x. One way to describe what we did is to say we added the coefficients. At first, it looks like those x's don't have coefficients. However, any term without a coefficient can be thought of as having a coefficient of 1. So it looks like we just added the coefficients: $1x + 1x = (1 + 1)x = 2x$.

More examples

$2x + 3x = (2 + 3)x = 5x$

$4x + 3x = (4 + 3)x = 7x$

Subtracting Algebraic Terms

$12x - 3x = (12 - 3)x = 9x$

$4x - 3x = (4 - 3)x = 1x = x$

The lesson here is: you can add or subtract coefficients when their terms are the same. Just to clarify, terms are only considered the same when they are to the same power. So 2x and 9x have similar terms, but $2x^3$ and 9x do not.

Multiplying Algebraic Terms

We can also multiply and divide terms. Again, let's take an example from arithmetic before we do the algebra. (The dot means multiply. Other ways of showing multiplication are (2)(2) and 2 X 2.)

$2 \square 2 = 2^2 = 4$

It might not be obvious here, but when you multiply a number by itself, you're really adding exponents.

Look again:

$2^1 \square 2^1 = 2^{1+1} = 2^2 = 4$

It is important to remember that when a number is written without an exponent, it is considered to be to the power of one. This is just like what we learned with variables without coefficients; we assume they have a coefficient of 1. So even though it isn't written, what is the power and exponent of x? 1 and 1. This is because, if we chose to write out the hidden numbers, $x = 1x^1$.

Now that we know this, we can multiply x and x.

$x \square x = x^2$

because

$x^1 \square x^1 = x^{1+1} = x^2$

We can take this quite a bit further.

First, notice that

$x \square x \square x = x^3$

This is because

$x^1 \square x^1 \square x^1 = x^{1+1+1} = x^3$

Now, let's multiply this by x

$x^3 \square x^{+1} = x^4$

Why?

Remember, we just add the exponents.

$x^3 \square x^1 = x^{3+1} = x^4$

The lesson here is: when multiplying terms with the same base, add the exponents. (In our examples, x is the base.)

Dividing Algebraic Terms

You can also divide algebraic terms. If you think about it, you might even be able to predict how we're going to do this. If we added exponents when we multiplied numbers with the same base, what do you think we're going to do when we divide numbers with the same bases?

We're going to subtract the exponents.

For example:

$x^5 / x^3 = x^2$

This is because

$x^5 / x^3 = x^{5-3} = x^2$

So when multiplying terms with the same base, add the exponents. When dividing terms with the same base, subtract exponents.

More on Multiplying and Dividing Algebraic Terms

Now, what would we do when multiplying numbers with the same base and different coefficients?

For Example:

$2x \square 3x$

First, let's scribble in those hidden exponents so we remember to add them. Now the problem looks like this:

$2x^1 \square 3x^1$

To make things a little easier, let's break up those terms to remind us that $2x^1$ means $2 \square x^1$, and $3x^1$ means $3 \square x^1$. Now we get:

$2 \square x^1 \square 3 \square x^1$

Okay. We're really close to finishing this problem. Do you remember the commutative property of multiplication? If you don't, I don't blame you. The commutative property of multiplication states that with multiplication, order doesn't matter. This means that we can rearrange this problem so that it looks like:

$2 \square 3 \square x^1 \square x^1$

You already know how to do $2 \square 3$ and $x^1 \square x^1$, right? $2 \square 3 = 6$ and $x^1 \square x^1 = x^{1+1} = x^2$. So now, we're left with:

$6 \square x^2$

The rest is easy, $6 \square x^2 = 6x^2$.

So to recap, what did we just learn? When we multiply algebraic terms, we multiply the coefficients. The same rule applies for division.

For example:

$4x^5 / 2x =$

well, first, let's write in the hidden exponent

$4x^5 / 2x^1$

Now, we can just divide 4 by 2 and x^5 by x^1.

$4/2 = 2$

and if you remember what to do when you divide terms with the same base:

$x^5 / x^1 = x^{5-1} = x^4$

Our last step is just to put it all together.

$4x^5 / 2x = 2x^4$

Algebraic Terms Raised to Powers

We have all the tools we need to add, subtract, multiply and divide algebraic terms. We just have one more related topic to cover: raising algebraic terms to powers and finding the square roots of algebraic terms. Yes, I saved the best for last. And, by now, you should know that this is nothing to fear. Raising algebraic terms to powers and finding the square roots of algebraic terms is, well, not as easy as $1 + 1 = 2$, but it can be pretty darn close.

Let's see:

$(x^2)^3 =$

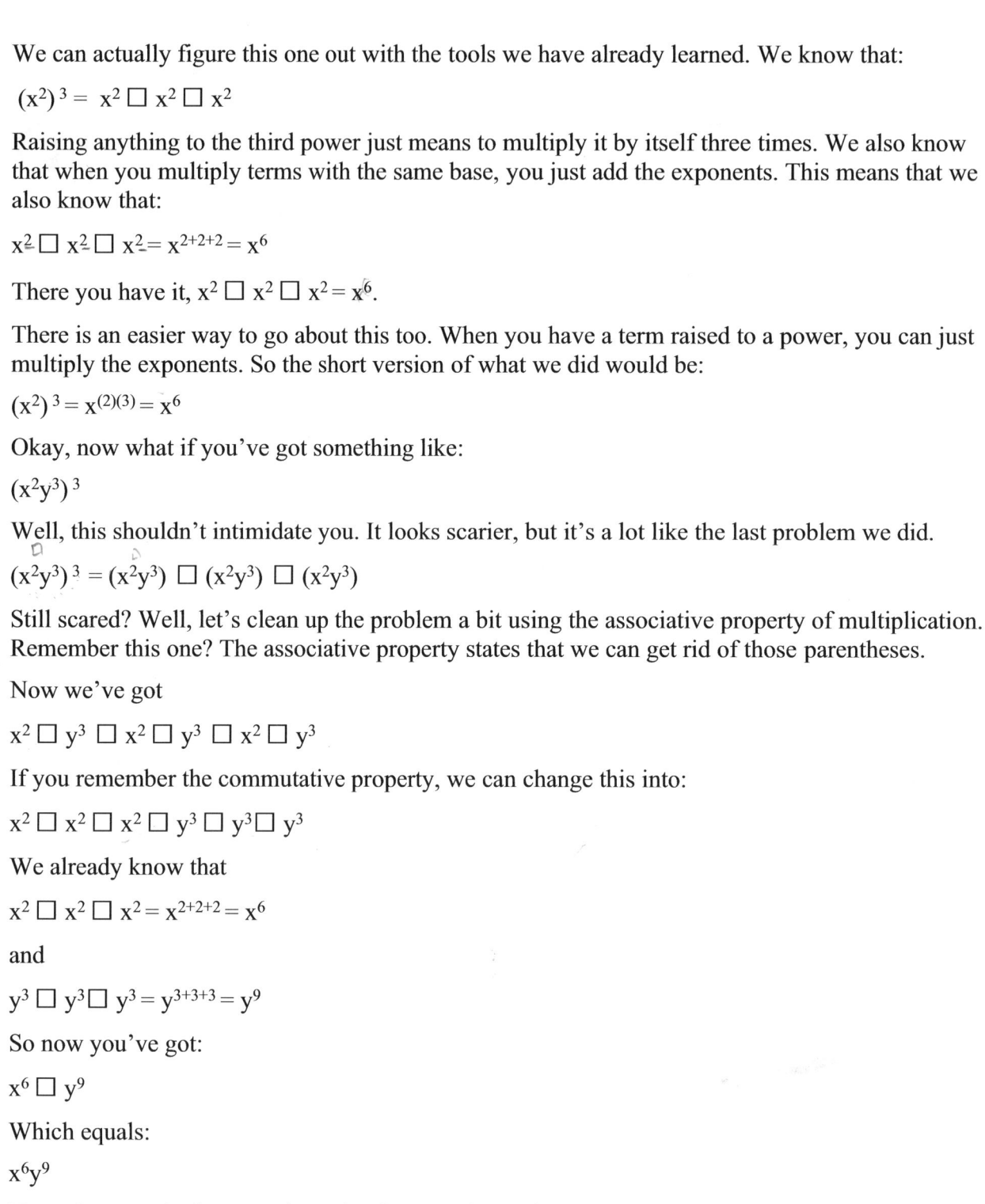

We can actually figure this one out with the tools we have already learned. We know that:

$(x^2)^3 = x^2 \square x^2 \square x^2$

Raising anything to the third power just means to multiply it by itself three times. We also know that when you multiply terms with the same base, you just add the exponents. This means that we also know that:

$x^2 \square x^2 \square x^2 = x^{2+2+2} = x^6$

There you have it, $x^2 \square x^2 \square x^2 = x^6$.

There is an easier way to go about this too. When you have a term raised to a power, you can just multiply the exponents. So the short version of what we did would be:

$(x^2)^3 = x^{(2)(3)} = x^6$

Okay, now what if you've got something like:

$(x^2y^3)^3$

Well, this shouldn't intimidate you. It looks scarier, but it's a lot like the last problem we did.

$(x^2y^3)^3 = (x^2y^3) \square (x^2y^3) \square (x^2y^3)$

Still scared? Well, let's clean up the problem a bit using the associative property of multiplication. Remember this one? The associative property states that we can get rid of those parentheses.

Now we've got

$x^2 \square y^3 \square x^2 \square y^3 \square x^2 \square y^3$

If you remember the commutative property, we can change this into:

$x^2 \square x^2 \square x^2 \square y^3 \square y^3 \square y^3$

We already know that

$x^2 \square x^2 \square x^2 = x^{2+2+2} = x^6$

and

$y^3 \square y^3 \square y^3 = y^{3+3+3} = y^9$

So now you've got:

$x^6 \square y^9$

Which equals:

x^6y^9

Now, that was the long version. The short version, using our trick, is to multiply each of the exponents by the power the term is raised to:

$(x^2y^3)^3 = x^{(2)(3)}y^{(3)(3)} = x^6y^9$

As I said, it's not quite as easy as 1 + 1 = 2, but it's close.

Square Roots of Algebraic Terms

If we multiplied exponents when raising algebraic expressions to powers, what do you think we'll do when we find square roots? Divide!

For example, find the square root of:

x^4

To find the square root of an expression raised to a power, divide the power by 2.

$x^{4/2} = x^2$

And there you have it, the square root of x^4 is x^2.

Conclusion

We've learned what algebraic expressions are, how to talk about them, and—most importantly—how to work with them. As it turns out, carrying out operations with algebraic expressions can be fairly simple. You've just got to remind yourself that you're using the same rules of math you've been using for years, now you're just applying them in a slightly more abstract way. If you commit the rules and definitions we've discussed here to memory, you'll have a really good foundation for more complicated problems dealing with this new, abstract area of math.

Chapter 2: Equations and Inequalities

In this section, we're going to cover the basic concepts involved in understanding and solving algebraic equations and inequalities. Equations are just statements that say two expressions are equal. Equations can be anything from the obvious, like saying one dollar is equal to four quarters, to something more obscure, like $e^{i\pi} + 1 = 0$. Inequalities are closely related to equations. Inequalities are statements that say that one expression is more that the other, one expression is less than the other, one expression is more than or equal to the other, or one expression is less than or equal to the other. These concepts build on each other and use a lot of the same rules. Equations are a bit easier to understand, so we'll start there.

Equations

Since equations are just statements that say two expressions are equal, they are a really useful way to find out information we do not know. Say, for example, we don't know how long a table is but do know that it is one foot shorter than the 10-foot-wide wall it is next to. You may have already guessed the answer, but mathematically, we would say:

$x + 1 = 10$

Formally, to solve this equation, we need to do what is called isolating the variable (x is the variable here). This means that we want to get x all by itself on one side of the equal sign. Luckily for us, we can move that 1 to the other side of the equation.

In order to do this, think about what we would have to do to get rid of + 1. To get rid of + 1, you just need to take away 1, or, – 1. Now, for the equal sign to remain true, we have to do this to both sides of the equation. So the first step in solving this equation is:

$x + 1 - 1 = 10 - 1$

The next step to solving this problem is to do what mathematicians called simplification. All this means is that we carry out whatever math we can. Here, it means that we can subtract one from one on the left of the equal sign and that we can subtract one from 10 on the right of the equal sign. If we do this, we get:

$x = 9$

Now we're done: the desk is nine feet long.

What we just solved is one of the simplest kinds of equations: one-step equations. In one-step equations, we have to do only one thing to get to the answer. In this case, we just had to subtract one from both sides. Let's go through a couple other one-step equations for practice.

As a general rule with one-step equations, just undo what is being done to the variable.

Let's say we want to solve:

3x = 12

This means a number multiplied by 3 is 12. How do we undo multiplying? Divide. Remember, we have to do this to both sides of the equation.

(3x) ☐ 3 = (12) ☐ 3

Now we have to remember our basic algebra, what is 3x ☐ 3? x.

So

x = 4.

Now, let's try a problem with division.

x/3 = 4

How do we do the opposite of dividing x by 3? We multiply x by 3. Remember, do this to both sides of the equation.

(x/3) ☐ 3 = 4 ☐ 3

Before we move on to the next step, let's pause to carefully go over doing (x/3) ☐ 3.

First, let's rewrite 3 as a fraction. 3 is a whole number, and you can rewrite any whole number as a fraction by putting it over 1. So 3 = 3/1.

We still have to do

(x/3) ☐ (3/1)

When multiplying fractions, always multiply the numerators (the numbers on top) by each other and multiply the denominators (the bottom numbers) by each other.

Now we get

(x/3) ☐ (3/1) = (x ☐ 3)/(3 ☐ 1)

Now we have to simplify the top and the bottom of the fraction. To do this, we'll have to do x ☐ 3. To make this a bit easier, write in x's coefficient. Remember, any variable without a coefficient has a coefficient of 1. When you multiply algebraic terms like 1x by constants like 3, just multiply the numbers.

1x ☐ 3 = (1 ☐ 3)x = 3x

Remember, this is just the top of the fraction. So right now, we've got:

3x/3

We have already covered how to simplify this.

3x/3 = x

Okay, so now we've got one side of the equation done. What about the other side?

4 ☐ 3 = 12

So

x = 12

Math is best learned through practice, so try to solve these problems on your own. Write down all of the steps it takes to solve these them, even if you can do them in your head. Writing down steps is good practice for when you face harder problems that are more complicated. If you have trouble solving these, look at the examples we've covered for help.

Practice

x – 1 = 5

2x = 10

x/2 = 10

x + 1 = 5

Solutions and explanations

1. x – 1 = 5

First, we need to do the opposite of – 1. This means we're going to add 1. Remember, we must do this to both sides of the equation.

x – 1 +1 = 5 + 1

Then simplify both sides:

x = 6

2. 2x = 10

What is the opposite of multiplying x by 2? Dividing by 2.

2x/2 = 10/2

Remember how to simplify 2x/2? Just divide the numbers.

x = 5

3. x/2 = 10

To do the opposite of dividing x by 2, multiply both sides by 2.

x/2 □ 2 = 10 □ 2

Even though we did this in great detail, I'm going to quickly explain how to simplify (x/2) □ 2 because it is such an important concept to understand. Skip this if you got it the first time.

    x/2 □ 2 = x/2 □ 2/1 (Remember, 2 = 2/1)

    x/2 □ 2/1 = (x □ 2)/(2 □ 1) (Multiply numerators and denominators)

    (x □ 2)/(2 □ 1) = 2x/2 (To simplify x □ 2, we think of x as 1x and then

    multiply the coefficient—1—by 2. x □ 2 = 1x □ 2 = (1□ 2)x = 2x)

$2x/2 = x$ Done! Or, as they say in the math world, Q.E.D. (Q.E.D. is an abbreviation for *quad erat demonstrandum*—Latin for "done!")

When we simplify both sides, we get:

$x = 20$

4. $x + 1 = 5$

Subtract 1 from both sides

$x + 1 - 1 = 5 - 1$

Simplify

$x = 4$

Inequalities

As I said in the beginning, inequalities are statements that say that one expression is more than the other, one expression is less than the other, one expression is more than or equal to the other, or one expression is less than or equal to the other. There are symbols for these different kinds of statements.

$x > y$ means "x is greater than y"

$x < y$ means "x is less than y"

$x \;\square\; y$ means "x is greater than or equal to y"

$x \;\square\; y$ means "x is less than or equal to y"

Memorizing these four symbols can be really easy if you notice the pattern. The wedge opens towards the bigger number. Some people think it is helpful to think of inequalities as hungry alligators that want to eat the bigger number. Also, notice that whenever there is a line under the inequality symbol, it could also be equal.

To make sure we understand what these symbols mean, let's put them to use. Which of the four symbols could we put in the blank?

3 _ 4

We could put < or $\square$. We could put < because 3 is less than 4. But why can we put $\square$? We can put the less than or equal to sign because three is less than four. We can use this symbol when the number to the right is less than *or* equal to the sign on the right.

So which signs could we put in this blank?

4 _ 4

We could put $\square$ or $\square$. Since 4 = 4, it is true that 4 is less than or equal to four. It's also true that 4 is greater than or equal to four.

Which signs could we put here?

4 _ 3

☐ or >. This is because 4 is greater than 3. So we could say, "four is greater than three", or, "four is greater than or equal to three." Both statements are true.

Learning inequalities opens a new world of algebra. It allows us to solve problems with inequalities. Solving problems with inequalities is actually really simple—it's almost exactly the same as solving equations. There is only one small difference between the two, but we'll get to that a little bit later. For now, we'll focus on problems where solving an inequality has the same steps as solving equations.

Let's start with:

x + 1 > 4

What would you do if this had an equal sign instead of an inequality? You'd subtract 1 from both sides.

x + 1 – 1 > 4 – 1

Next, simplify both sides of the equation.

x > 3

Easy! However, some people get scared by that inequality. If that little wedge is throwing you off, get rid of it! A lot of people replace the inequality with an equal sign while they're working on the problem, then put it back in when they're done. If we did that, our problem would look like this:

x + 1 > 4 ☐ x + 1 = 4

     x + 1 = 4

     x + 1 – 1 = 4 – 1

     x = 3 ☐ x > 3

x > 3

So if you could solve any equation, you could solve any inequality. They're basically the same thing—except for that one difference I mentioned. Before we get to that, let's do one more practice problem to make sure we've got the basics down.

2x ☐ 4

First, we do the opposite of multiplying x by 2. This means, divide both sides by 2.

2x/2 ☐ 4/2

Simplify.

x ☐ 2

Okay, so we've got the basic idea down: inequalities are not that bad. But, what's that one difference between solving inequalities and solving equations? The one rule you need to memorize when working with inequalities is: **whenever you multiply or divide both sides of an inequality by a negative number, flip the inequality sign**. What does "flip" mean? It means < becomes >, > becomes <, ☐ becomes ☐, and ☐ becomes ☐.

Let's do some problems to show how that works.

-3x > 9

To solver this, first we would divide both sides by -3.

-3x/-3 > 9/-3

When we simplify, remember to replace > with <.

x < -3

This rule also applies to problems where we multiply both sides by a negative number:

x/-3 < 3

To solve this problem, we multiply both sides by -3

x/-3 ☐ -3 < 3 ☐ -3

Simplify and flip the sign

x > -9

Now, try to solve some inequalities on your own.

Practice

x – 1 ☐ 2

x + 1 ☐ 2

-2x > 4

x/2 > -4

Solutions and explanations

1. x – 1 ☐ 2

First, add 1 to both sides.

x – 1 + 1 ☐ 2 + 1

Simplify

x ☐ 3

2. x + 1 ☐ 2

Subtract 1 from both sides.

x + 1 – 1 □ 2 – 1

Simplify

x □ 1

3. -2x > 4

divide both sides by -2

-2x/-2 > 4/-2

Simplify and flip the sign because you are dividing both sides by a negative number.

x < -2

4. x/2 > -4

Multiply both sides by 2

x/2 □ 2 > -4 □ 2

Simplify

x > -8

Chapter 3: Relations & Functions

Relations

A relation, when we're talking math, is a collection of two groups of numbers or variables. Usually, when we talk about relations, these groups are named "x" and "$f(x)$". ($f(x)$ is pronounced "f of x"—it means function of x.)

Say I wanted to collect information about the relationship between someone's age and how fast they can run. One way to do this would be to make a table. In one column, I would write people's ages. In the other column, I could record how long it takes for them to run a mile. Here's an example of what that would look like:

Age	Mile Time
12	14 Min.
14	13 Min.
18	7 Min.
20	10 Min.
20	8 Min.
21	11 Min.
55	15 Min.
70	20 Min.
19	18 Min.

As you can see, the data from our running survey gave us two sets of numbers: people's ages and their mile times. This table is a perfect example of a relation. If we wanted, we could label these columns "x" and "$f(x)$". Then, the table would look like this:

x	$f(x)$
12	14
14	13
18	7
20	10
20	8
21	11
55	15
70	20

19   18

In any relation, there is a name for the groups of number. The first group, in this case the x group, is called the domain. The second group, in this case the $f(x)$ group, is called the range.

One of the most basic things you might get asked to do when dealing with relations is to find the domain and the range. When answering one of these questions, all you have to do is write down all the values in that group in order without repeating numbers.

For example, we said that the x group was the domain. If you were asked to write the domain of our data about running, you would put:

Domain: {12, 14, 18, 19, 20, 21, 55, 70}

Notice that in the list we started with, 20 was repeated twice and 19 came after 70. However, when we answer a question about the domain of our data, we get rid of the extra 20 and put 19 in the right place.

The same rules apply for finding the range of our data. We put the values in order without repeating. What is the range of our data?

Range: {7, 8, 10, 11, 13, 14, 15, 18, 20}

Besides the fact that there are some new, unfamiliar words being used here, this should seem simple enough. So simple, in fact, that you're probably asking why we're even talking about relations, domain, and range. We're developing these concepts so we can talk about a special kind of relation: functions.

Functions

Functions are relations that have exactly one value in the range for each value in the domain. If that definition sounds a like a bunch of gibberish, don't worry; we're going to work on smoothing things out. It's really important to understand the difference between relations that are functions and relations that are not functions.

So was the data about running a function or just a relation? If you look back at the definition of function, it says that functions have one value in the range for each value in the domain. What does this mean? It means that there should be only one range per domain.

To help with our understanding of this problem, let's copy the table. This time, however, we'll write the values from the domain in order without repeating. We'll list their matching relations next to them.

x (Domain)	$f(x)$ (Range)
12	14
14	13
18	7

19	18
20	10
	8
21	11
55	15
70	20

If you look at our list, the number 20 has two corresponding values in the range: 10 and 8. That means that our list is not a function. So what does a function look like?

Let's say that I make a number generator. The generator takes whatever number I plug in, adds 2, and then gives me a new number. For example, if I plugged in 2, the program would add 2 and 2 and spit out 4. Since we're starting with the numbers we plug in, that will be our domain. The numbers that come out will be our range. Here's a table showing the data we get for some random numbers I plug into the generator:

Domain	Range
2	4
6	8
4	6
7	9
9	11

If you notice, each number in the range matches up to exactly one number in the range. 2 only matches up with 4, 6 with 8, and so on.

Say I plugged 2 into the generator twice and my data was:

Domain	Range
2	4
2	4
6	8
4	6
7	9
9	11

Is this still a function? Well, let's see. What's the domain?

Domain: {2,4,6,7,9}

How about the range?

Range: {4, 6, 8, 9, 11}

What would we see if we rewrote the table so that we listed values from the domain in order without repeating and then listed the matching ranges?

We would get:

Domain	Range
2	4
4	6
6	8
7	9
9	11

This is a function.

For practice, let's see if you can identify the following relations as functions or merely relations:

1.
Domain	Range
3	1
3	2
3	3

2.
Domain	Range
1	3
2	3
3	3

Solutions and Explanations

1. Merely a relation

Why? If you're having trouble with this, answer this question: what is the domain?

Domain: {3}

Now, recreate the table starting with the domain.

Domain	Range
3	1
	2
	3

As you can see, there is more than one range for the value in the domain. This means that this is not a function—it is merely a relation.

2. Function

If you are unclear as to why, let's start with the domain.

Domain: {1, 2, 3}

Recreate the table starting with the domain:

Domain	Range
1	3
2	3
3	3

Now ask yourself, does this have exactly one value in the range for each value in the domain? Yes! And it's okay that they are all the same value. This *is* a function.

Function Rules

Remember the number generator that helped us come up our function? One way to describe the generator is to say that it followed a rule: add 2. Since that rule helped us come up with a function, we could say that this rule is a function rule, or, an equation that describes a function.

Since function rules are equations that describe functions, we should be able to make a function based off of a function rule, and *vice versa*. Let's start by making a function based off of a function rule.

This time, we're going to write our function rule formally. In order to do this, we need to learn function notation. Function notation just means that we're going to write the rule describing how the two sets of numbers in our relation, the domain and range, relate to each other in the form of an equation.

Before we can write a function rule in function notation, we need to remember what x and $f(x)$ are. x refers to the domain; $f(x)$ refers to the range. So if I wanted to write a really simple equation where the domain is the same as the range, I could write $f(x) = x$. Now we can make our table.

To make a table, pick some numbers that you want to plug into the function rule. When I make a table, I usually pick at least three numbers: a negative number, a positive number, and zero. So for our domain, let's use 1, 0, and -1.

What are our ranges going to be?

When you plug in your domain values, you just put them where x is in the function rule. For example, if our rule is $f(x) = x$ and we want to find $f(x)$ when x = 0, we could write:

$f(0) = 0$

If x = 1, we would write:

$f(1) = 1$

And if x = -1, we write:

$f(-1) = -1$

Now we can fill out a table. The x values will be the input, the $f(x)$ values will be the output.

x	$f(x)$
0	0
1	1
-1	-1

Now, let's try to construct a table for a function that's a bit more complicated.

$f(x) = -x + 5$

First, decide what your domain (the number you will plug in for x) is going to be. I think 0, 1, and -1 are good candidates for the job. Let's see what you get when you plug those values into our function rule.

What does $f(x)$ equal if x = 0?

First, put 0 in for x:

$f(0) = -(0) + 5$

Then simplify.

$f(0) = -0 + 5 = 5$

$f(0) = 5$

So when x = 0, $f(x)$ 5.

Next, plug in 1.

$f(1) = -(1) + 5$

Simplify.

$f(1) = -1 + 5 = 4$

Finally, plug in -1

$f(-1) = -(-1) + 5$

$f(-1) = 1 + 5 = 6$

$f(-1) = 6$

When you are done with your calculations, organize your values in a table:

x	$f(x)$
0	5
1	4

-1   6

Try to make a table for $f(x) = -3x$ on your own.

Solution and Explanation

To make a table for $f(x) = -3x$, first choose which number you are going to plug in for x. You can choose any numbers you want, but in this example I'm going to work with—you guessed it—0, 1, and -1.

Let's go ahead and plug those numbers in.

$f(0) = -3(0)$

$f(0) = 0$

So when $x = 0$, $f(x) = 0$.

$f(1) = -3(1)$

$f(1) = -3$

So when $x = 1$, $f(x) = -3$.

$f(-1) = -3(-1)$

$f(-1) = 3$

So when $x = -1$, $f(x) = 3$.

Now, organize that information:

x	f(x)
0	0
1	-3
-1	3

Writing Function Rules

The other side of function rules is looking at functions and coming up with the function rule that describes them. Before trying to come up with a function rule for a function, it is very important to make sure that the groups of numbers you are working with are, in fact, functions.

Let's start with a simple example: write a function rule for the following function.

x	f(x)
0	1
1	2

-1   0

First, ask yourself, "Is this a function?"

To answer this question, find the domain. Remember, the domain is the first set of numbers; in this case it is the collection of x values.

Domain: {-1, 0, 1}

Then, rewrite the table accordingly.

x	f(x)
-1	0
0	1
1	2

Does this have exactly one value in the range for each value in the domain?

Yes—so it's a function.

Now we can work on writing a function rule. Since the numbers are now in order, it might be easier to see what the relationship is between the x values and the $f(x)$ values.

A good place to start is the first row of our table, when $x = -1$ $f(x) = 0$. What can you do to -1 to get 0? You could add 1. Maybe the function rule is $f(x)$ is equal to x plus 1, or, $f(x) = x + 1$.

We can test this hypothesis to see if $f(x) = x + 1$ describes what is going on in other rows. Does $f(x) = 1$ when $x = 0$? Let's see.

$f(0) = 0 + 1$

$f(0) = 1$

Good, $f(x)$ *does* equal 1 when x equals 0.

For the sake of completeness, let's test our hypothesis against that last row. So now we ask, "Does $f(x) = 2$ when $x = 1$?"

$f(1) = 1 + 1$

$f(1) = 2$

Yes! We've got the right function rule: $f(x) = x + 1$.

Now, try to come up with a function rule for the following relation on your own:

x	f(x)
-1	-1
0	1
1	3

Solution and Explanation

The function rule for the table is f(x) = 2x + 1

Before trying to solve this, establish that the groups of numbers on the table form a function. In this case, determining this is easy. The domain is in order and doesn't repeat numbers. This is a function.

Next, start with a row and think of possible function rules. In the first row we have f(x) = -1 when x = -1. There are many possible rules here. Here are a few:

f(x) = x

f(x) = 2x + 1

f(x) = 3x + 2

Test these against the other rows. You'll see that the only function rule that applies to all three rows is f(x) = 2x + 1.

Chapter 4: Polynomial and Rational Functions

Polynomial Functions

Polynomial functions are nothing to be scared of. They are just functions that are relations that have exactly one value in the range for each value in the domain. These functions involve using polynomials or algebraic expressions with more than one term. If the terms function, domain, polynomial, or algebraic expression sound unfamiliar, go back and read the sections on those topics.

That might sound bad, but working with polynomials can be very easy. Polynomial functions can include functions as simple as $f(x) = x + 1$, $x = 1$. They can also include problems as such as $f(x) = x^2$, $x = x^2 + 3x$. Luckily, by the end of this section, both of these problems should be a piece of cake.

Before we get to working with polynomial functions, we need to introduce some vocabulary.

Standard Form

First and foremost, it is important to understand standard form. When a polynomial is in standard form, the terms are arranged in order of descending degree. This sounds a bit complicated, but it is actually very easy to put any polynomial into standard form.

For example:

$4x^2 - 3x - 2 + 5x^3$ is not in standard form.

If we are going to put this polynomial into standard form, we need to put the term raised to the highest power first. Can you guess which one that is?

If you guessed $5x^3$, you're right. All you have to do is move that term to the front of the polynomial and it will be in standard form.

$5x^3 + 4x^2 - 3x - 2$

Now that we understand standard form, we need to discuss two important concepts. First, when moving terms around in a polynomial we must move the sign along with it.

For example:

$4x^3 - 6x - 1 - 5x^2$ is not in standard form.

In this polynomial, $-5x^2$ needs to be moved between $4x^3$ and $-6x$. Now we get:

$4x^3 - 5x^2 - 6x - 1$

The other important concept answers the question, "why does -1 come last?"

The key to understanding this is to realize that $-1 = -1x^0$. This is because anything raised to the power of zero is always one. So we think of constants, like -1, as algebraic terms that consist of constants multiplied by a variable that happens to equal 1. So any constant in a polynomial is considered to be an algebraic expression raised to the power of zero. That way, when we write in the assumed details, we get the powers in descending orders. (Remember, $x = x^1$.)

$4x^3 - 5x^2 - 6x^1 - 1x^0$

Naming Polynomials

We've just explained that it is standard to arrange polynomials in a way that puts the term raised to the highest power first. This is partly because polynomials are named after the degree of their highest term:

Name	Highest Power	Example
Constant	0	1 (Remember that $1 = 1x^0$)
Linear	1	$x + 1$ (Remember that $x = x^1$)
Quadratic	2	$x^2 + x + 1$
Cubic	3	$x^3 + x^2 + x + 1$
Quartic	4	$x^4 + x^3 + x^2 + x + 1$
Quintic	5	$x^5 + x^4 + x^3 + x^2 + x + 1$

Working with Polynomials

When working with polynomial functions, it is very important to know how to multiply, divide, add, and subtract polynomials. A good deal of the key concepts behind carrying out these operations was covered earlier under the section on algebraic expressions. Here we will cover a couple of good rules to know when working with polynomials.

Multiplying polynomials can be simple. Don't believe me? Try to do this problem:

```
    56
  X 41
    56
+ 2,240
  2,296
```

What does this have to do with multiplying polynomials? Everything! The trick to algebra is being able to carry out the operations of arithmetic with unknown quantities. Let's look at this problem again, paying close attention to the logic behind it, and see if it helps us understand how to multiply polynomials.

The first step in understanding how this applies to polynomials is to think of 41 as 40 and 1, so—without changing the problem—I'm going to replace 41 with 40 + 1.

    56

    X 40 + 1

    56   To get 56, multiply 1 and 56.

    + 2,240   To get 2,240, I multiply 40 and 56.

    2,296   To get the final answer, add everything up.

Now let's try to apply this same logic to a problem with polynomials. Let's multiply (y − 1) by (y + 1). (To avoid confusion, I'm not going to write the X for multiplication, because it looks too much like a variable.)

    y − 1

    y + 1

    y − 1   Multiply 1 and y − 1

      $y^2 - y$ Multiply y and y − 1. To do this, multiply − 1 and y (which equals −y), then multiply y by y (which equals y^2).

Now, just like with 56 x 41, combine the terms.

    y − 1

+ $y^2 - y$ Notice that I lined up like terms to make this step easier.

    $y^2 - 1$

Putting it all together

Try to answer the following question. A full explanation will be provided. If you have trouble answering this question, review this section, the section on algebraic expressions, and the section on relations and functions.

Practice:

Evaluate $f(x) = x \,\square\, (y + 3)$ when $x = (y + 2)$. Write your solution in standard form and name the polynomial based on its degree.

Answer and Explanation:

Answer: $y^2 + 5y + 6$; quadratic.

Explanation:

Plug in y + 2 where you see x in $f(x) = x \cdot (y + 3)$. Now you should have:

$f(y + 2) = (y + 2) \cdot (y + 3)$

Now, multiply

$$\begin{array}{r} y + 2 \\ \underline{y + 3} \\ 3y + 6 \quad (3 \text{ times } y + 2 = 3y + 6) \\ \underline{y^2 + 2y} \, (y \text{ times } y + 2 = y^2 + 2y) \\ y^2 + 5y + 6 \end{array}$$

This is already in standard form. The highest degree is 2, so the expression is quadratic.

Rational Functions

Rational functions are functions that can be described as the result of the division of polynomials.

That might sound tough, but rational functions can be as simple as $f(x) = 1/x$. However, rational functions can also include more complicated problems. In this section, we'll cover the important skills you'll need to develop in order to work with these functions.

Before we get into the math, it is important to note that functions with variables below the fraction bar, like $f(x) = 1/x$, present a problem that we have not yet dealt with. What would happen if x = 0?

Well, it is impossible to divide by zero; so we just can't have x = 0. We represent this by writing $f(x) = 1/x$, where $x \neq 0$.

This seems simple enough, but we must remember that it isn't always true that $x \neq 0$. Consider the function $f(x) = 1/(x + 1)$.

In this function, x can be 0 because:

$f(0) = 1/(0 + 1)$

$f(0) = 1/(1)$

$f(0) = 1$

However, $x \neq -1$ because:

$f(0) = 1/(-1 + 1)$

$f(0) = 1/(0)$

$f(0) = \emptyset$

Ø means there is no solution.

In every rational function that has a variable in the denominator, there is a number that the number cannot equal. This is important to keep in mind when working with rational functions.

Identify the value that x cannot be in the following functions:

1. $f(x) = (x^2 - 4)/x$
2. $f(x) = x/(x^2 - 4)$
3. $f(x) = x/(x - 3)$

Solutions and Explanations

1. $x \neq 0$

This is because the denominator is x and we cannot have a zero as a denominator.

2. $x \neq 2$ or -2

This is one is a bit more difficult. If you factor out $x^2 - 4$, the answer is a bit more obvious.

$x^2 - 4 = (x - 2) \square (x + 2)$

To check your factorization, you can always multiply. (You can use the method shown in the section on polynomial functions)

$$\begin{array}{ll} x - 2 & \\ \underline{x + 2} & \\ 2x - 4 & \text{Get this by multiplying } x - 2 \text{ by } 2. \\ \underline{x^2 - 2x} & \text{Then multiply x and } x - 2. \\ x^2 \quad\quad - 4 & \text{Combine } 2x - 4 \text{ and } x^2 - 2x. \end{array}$$

Now, x cannot be 2 because you would have $(2 - 2) \square (2 + 2)$ in the denominator. This is a problem because $2 - 2 = 0$, and $0 \square (2 + 2) = 0$.

The same logic applies when $x = -2$

3. $x \neq 3$

This is because $3 - 3 = 0$ and we cannot have 0 as a denominator.

□ Multiplying and dividing rational expressions

When working with rational functions, it is very important to be able to multiply and divide rational expressions.

When multiplying fractions, remember to multiply the numerators by the numerators and the denominators by the denominators. For example, when you multiply 1/3 and 1/5:

$1/3 \cdot 1/5 = (1 \cdot 1) / (3 \cdot 5) = 1/15$

The same goes for rational expressions.

$x/y \cdot x/y = (x \cdot x) / (y \cdot y) = x^2 / y^2$

Dividing fractions is not much more difficult than multiplying fractions. In case you don't remember, this is how you divide fractions:

(1/3) / (1/5)

Flip the denominator (that's the fraction on the bottom) and multiply.

(1/3) • (5/1)

The rest is easy.

(1/3) • (5/1) = (1 • 5) / (3 •1) = 5/3

Now let's try this with some variables.

(x/y) / (2/3)

Flip the denominator and multiply.

(x/y) • (3/2) = (3x/2y)

Adding and subtracting rational expressions

In order to work with rational functions, it will also be necessary to be able to carry out addition and subtraction with rational expressions. Luckily, these two concepts are closely related and are similar to arithmetic you already know. Consider the following problem:

1/3 + 1/5

In order to carry out this operation, the fractions must have common denominators, so we multiply the fractions by a number that will give us common denominators.

In this case we need to multiply 1/3 by 5/5, and we need to multiply 1/5 by 3/3.

This gives us:

5/15 + 3/15

Now, we add the numerators:

5/15 + 3/15 = 5+3/15 = 8/15.

The same concept applies to rational functions.

To illustrate this, let's try to do 1/x + 1/y.

In order to do this, what do we need to multiply 1/x by? We need a common denominator. xy seems like it will work for us, so multiply 1/x by y/y and 1/y by x/x to get common denominators.

As we have seen in the section on multiplying rational expressions:

(1/x) • (y/y) = y/xy

and

(1/y) • (x/x) = x/xy

now, we can add the two expressions

(y/xy) + (x/xy) = (y + x)/xy

Practice

$f(x) = (y + 2) / (x)$

1. What can x not be? Why?

2. What is $f(x)$ if $x = (y + 2) / y$?

Answers and Explanations

1. $x \neq 0$

This is because it is impossible to divide by zero.

2. If we plug (y + 2) / y in for x we get:

$f((y + 2) / y) = (y + 2) / ((y + 2) / y)$

This might look scary, but you have all the tools necessary to solve this problem.

We have a fraction in the denominator, so we know we're going to flip it and multiply. To make this easier, we'll express y + 2 as a fraction.

(y + 2) = (y + 2) / 1

So we get:

(y + 2) / ((y + 2) / y) = ((y + 2) / 1) / ((y + 2) / y) = ((y + 2) / 1) • ((y / y + 2)

Now, we multiply the numerators and denominators:

((y + 2) / 1) • ((y / y + 2) = (y • (y + 2)) / (1 • (y + 2)) = $(y^2 + 2y) / (y + 2)$

It might look like you are done, but there is one more step that can be done here:

$y^2 + 2y = y • (y + 2)$

and

(y • (y + 2)) / (y + 2) = y

Chapter 5: Systems of Equations

When you have an equation like x = y, you can't really solve it for x or y. This is because the equation has two variables. When we see an equation like this, the solution is a table of numbers or a line on a graph that goes on forever. The table of numbers and the line show that for every different value of x, there is a different value for y.

When you have two equations with two variables, however, you can sometimes find a unique answer that solves both equations. Problems where you have multiple equations that you use to find a unique answer are called systems of equations. A typical systems of equations problem might look something like this:

10 = x + y

2x + y = 20

Solve for x and y

There are several ways to solve systems of equations problems. Three of them are:

☐ Substitution

☐ Elimination

☐ Graphing

Graphing

In this tutorial, we're going to start with the graphing method. This is because graphing systems of equations lets you see how we get one solution for the two problems. This will help us understand how substitution and elimination work.

To solve a system of equations by graphing, you transform the equations so both begin with y =. Transforming equations is exactly like solving equations. The only difference is that when you are done you won't be left with a number as an answer. Instead, you will end up with an algebraic expression. For this problem, we would start with:

10 = x + y

Then, move y to the outside.

10   = x + y

 − y   − y   Subtract y from both sides of the equation

10 − y = x

Now, move 10 to the other side of the equal sign.

10 − y = x

–10   – 10   Subtract 10 from both sides of the equation

-y = x – 10

Now, we need to get rid of the negative sign attached to y. To do this, remember that -y = -1y. And that -1 □ -1 y = y. This means that you can get rid of negative signs by simply multiplying both sides of the equation by -1.

So

-y = x – 10 is the same as -1y = x – 10

Multiply both sides by -1 to make y positive.

-1 □ (-1y) = -1 □ (x -10)

When performing the operation -1 □ (x -10), remember to use the distributive property—multiply x and -10 by -1.

Now you should have

y = -x + 10

Use this to make a table. Do this by plugging in values for x and finding y. Good values for x are 1, 0 and -1.

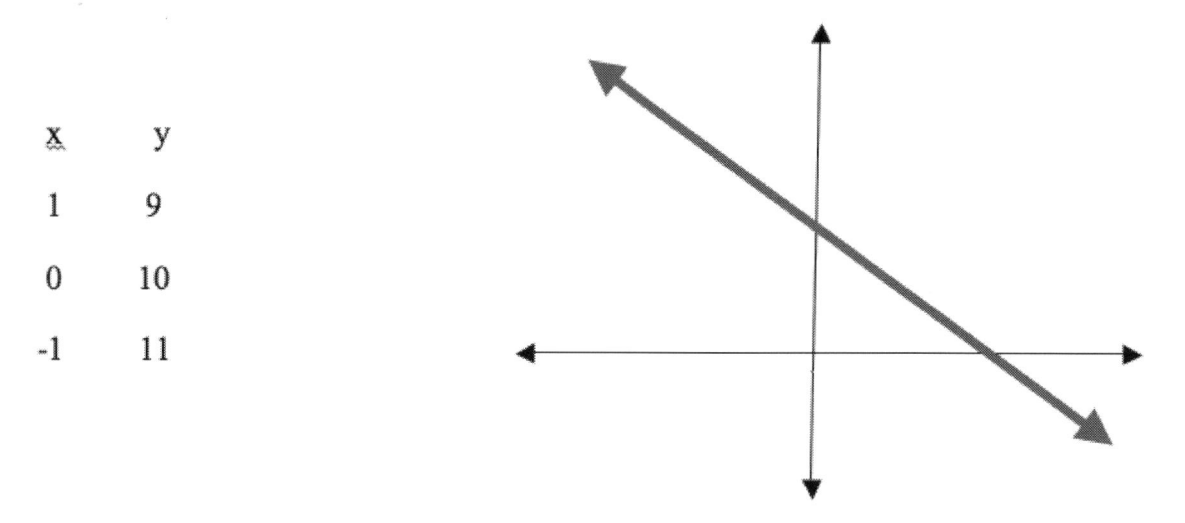

x	y
1	9
0	10
-1	11

Use this to create a graph.

Use similar steps to create a graph for 2x + y = 20.

Begin by moving 2x to the other side of the equation.

$$2x + y = 20$$
$$\underline{-2x \qquad\qquad -2x}$$
$$y = 20 - 2x$$

Create a table:

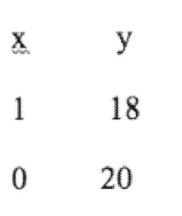

x	y
1	18
0	20
-1	22

If you graphed both equations on the same plane you would get:

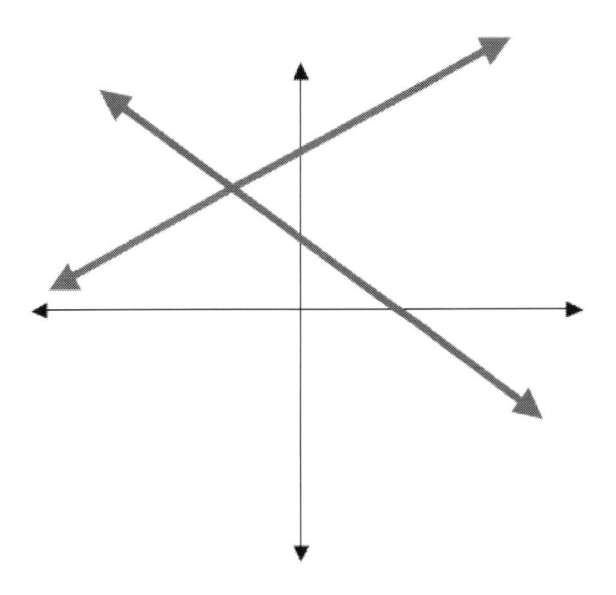

As you can see, these lines intersect at one point. The coordinates of this point give us the unique solution to the system of equations. As you can see, however, it is difficult to tell the exact coordinates at which they intersect. This is one of the problems with the graphing method. This is the motivation behind using the next two strategies we will cover.

For the record, on this graph the coordinates of the point of intersection is (-1, 10). Since coordinates are written (x, y) that means that the solution to the system is x = -1, y = 10.

Substitution

When you use the substitution method to solve systems of equations, you get exact numerical solutions. This makes substitution a very useful strategy. To use substitution, you solve one of the equations for x or y. We did this already in our example on graphing. We solved both equations for y. Next, you plug one equation directly into the other. Like this:

y = -x + 10

y = 20-2x

Let's take y = -x + 10 and plug it in to y = 20-2x

$$y = -x + 10 \qquad y = 20 - 2x$$

$$-x + 10 = 20 - 2x$$

Now we just have to solve for x.

-x + 10 = 20 – 2x

Get x onto only one side of the equal sign.

–x + 10 = 20 – 2x

+x   + x Add x to both sides of the equation

   10 = 20 – x

Isolate x

10 = 20 – x

-20 -20   Subtract 20 from both sides

-10 = -x

Remember, -x = -1x and that we can get rid of -1 by multiplying both sides by -1.

-1 ☐ -10 = -1 ☐ -1x

10 = x

The rest is easy. You just plug 10 in for x in one of the equations, then solve for y.

Let's plug x into y = 20 – 2x

y = 20-2(10)

Simplify

y = 20 – 20

y = 0

The solution to the system is x = 10, and y = 0. Since coordinates are written (x, y), the place where these two lines would cross on a graph is (10, 0).

Elimination

We could solve systems of equations a third way. This method, like the substitution method, delivers exact answers. However, as you will see, it can be used only part of the time. This is because elimination works a lot like finding common factors; if there are no common factors, the strategy will not work. Let's take a look at an example to see why.

y = -x + 10

y = 20 – 2x

When we use the elimination method, we need to have like terms with the same coefficients in both equations. If that sentence just went right over your head, don't worry! This just means that you want the same number in front of x or y. In this case, we have this. We have no coefficient in front of y. So this step is already done for us. Next we're going to subtract one equation from the other. This might sounds strange, so it might make more sense if we take a look at an example. We're going to subtract y = 20 – 2x from y = -x + 10. Before we do this, we need to rearrange the terms so that they line up neatly when you do the subtraction step.

First, the rearrangement:

y = 20 – 2x is equal to y = -2x + 20. Now you can subtract.

   y = -2x + 20

-(y = -x + 10)

   0 = -x + 10

Next, solve for x.

0 = -x + 10

Isolate x:

0 = -x + 10

-10   - 10

-10 = -x

To get rid of the negative sign, multiply both sides by -1.

-1 □ -10 = -1 □ -x

10 = x

To solve for y, do exactly what you did in the second part of the substitution method: plug in 10 for x and solve for y.

We already know that this will give us 0.

As you can see, systems of equations can be solved all three ways. Many students feel overwhelmed by this. My suggestion to those students is to work towards understanding one of these methods very well and remembering that the other two are out there.

When picking a method to focus on, my advice is to learn the substitution method. Of all three methods presented here, this is the only one that delivers exact answers and works with any equation. The downfall of the graphing method is that your coordinates could land between whole numbers, making it difficult to find the correct answer. The problem with the elimination method is that finding compatible coefficients could be very difficult. For the sake of practice, try solving the next system of equations.

-2x + 3y = 8
3x - y = -5

Answer and explanation:

3x - y = -5

-2x + 3y = 8

Elimination:

Find which variable you want to eliminate. It seems that eliminating y would be easiest. To do this, you will need to multiply both sides of 3x - y = -5 by 3.

3 □ (3x – y) = 3 □ (-5)

9x – 3x = -15

Now, combine this equation with -2x + 3y = 8 to eliminate 3y.

-2x + 3y = 8

9x – 3x = -15

7x   = -7

Now, solve for x.

7x = -7

Isolate x by dividing both sides by 7.

7x/7 = -7/7

x = -1

Next, plug -1 in for x and solve for y.

-2(-1) + 3y = 8

2 + 3y = 8

Subtract 2 from both sides

  2 + 3y = 8

-2   -2

     3y = 6

Solve for y.

3y/3=6/3

y = 2.

Substitution:

3x - y = -5

-2x + 3y = 8

Solve for y in one of the equations.

3x – y = -5

-3x   -3x

  - y = -5 – 3x

-1(-y) = -1(-5 – 3x)

y = 5 + 3x

Plug y = 5 + 3x into -2x + 3y = 8

-2x + 3(5 + 3x) = 8

-2x + 15 + 9x = 8

15 + 7x = 8 Combine like terms (-2x and 9x)

15 + 7x = 8

-15   -15

     7x = -7

7x/7 = -7/7

x = -1

As we saw in the last explanation, from here you plug in -1 and get y = 2.

Graphing:

If you graph both equations, you get a graph that looks like this:

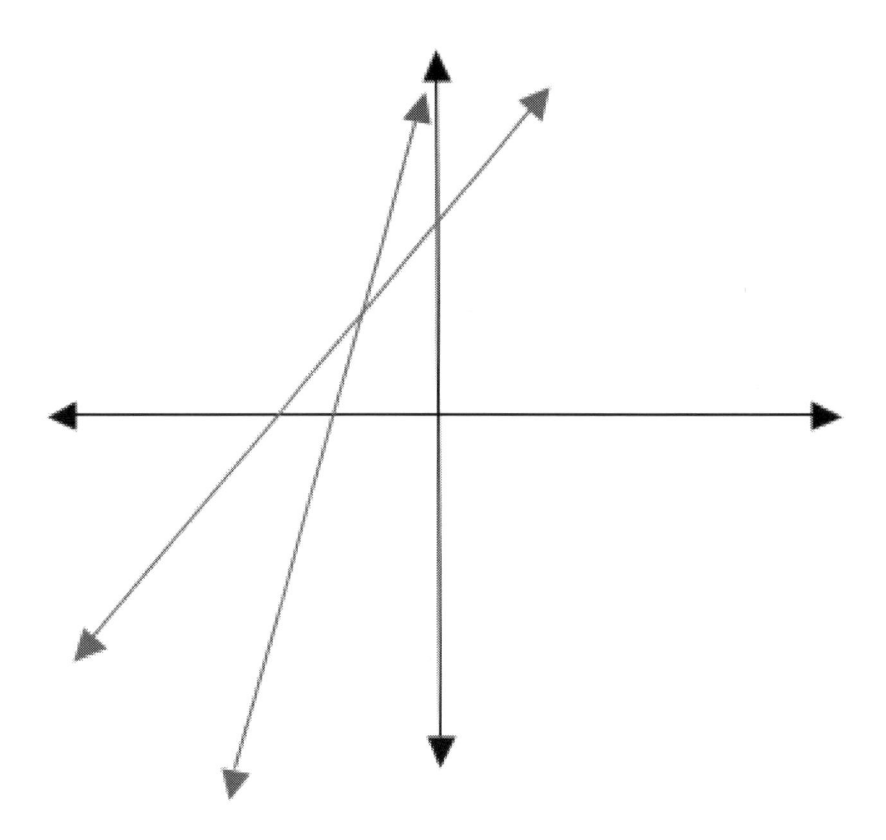

The red line is $3x - y = -5$ (which is the same as $y = 5 + 3x$)

The blue line is $-2x + 3y = 8$ (which is the same as $y = (8+2x)/3$)

Systems of inequalities

Solving systems of inequalities is a lot like solving systems of equations with graphing. The only difference is that when you are doing systems of inequalities, you have to remember to follow the inequality rules.

Inequality review:

< means less than, > means greater than, ≤ means less than or equal to, and ≥ means greater than or equal to.

Whenever you multiply or divide both sides of an inequality by a negative number, you must reverse the direction of the inequality. For example, < becomes >, > becomes <, and so on.

When you graph inequalities, inequalities with ≤ and ≥ get solid lines; inequalities with > and < get dashed lines.

When you graph an inequality, if it is less than y (y > or y ≥), you color above the line. If the inequality is greater than y (y <, or y>), color below the line.

Let's try to graph an inequality

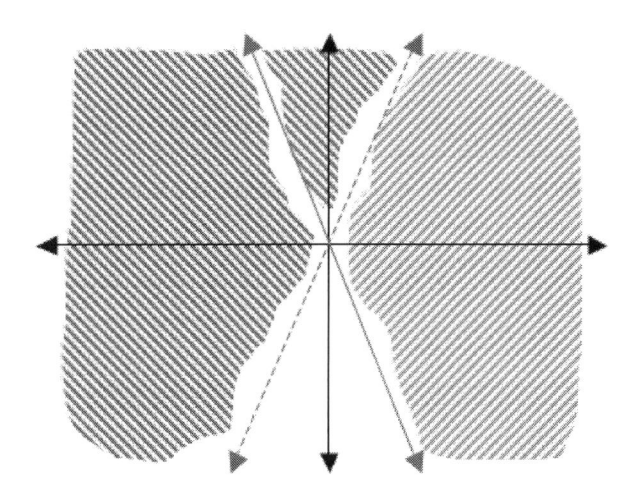

$y > -x$

$y \geq x$

Graphing:

Graph both inequalities,

remember to shade accordingly.

The purple area is where the

shaded areas overlap.

The purple area is known as the solution region. The solution region is what we are looking for when we solve systems of inequalities. Any point that you choose in this region will be a solution to the system.

For example, if you chose (0, 10) this would be a solution to both equations because 10 > 0 and 10 ≥ 0.

If we show only the solution region, it looks like this:

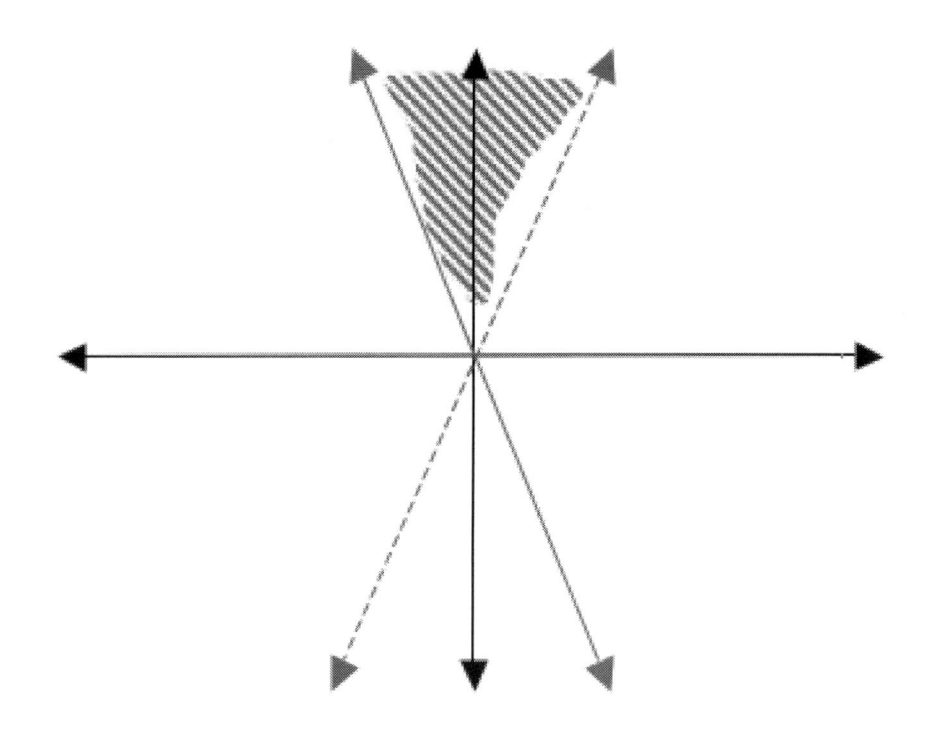

Conclusion

We covered a lot in this section! This should be fine, though. Solving and graphing systems of inequalities and equations requires many skills. This is why this is the last section of the algebra guide—to do these types of problems, you need to understand all the basic algebra concepts we've covered so far.

If you find yourself having trouble with any of this, review the other sections of this guide. Sometimes a problem with these types of problems comes from a misunderstanding of a more basic concept—like understanding algebraic expressions or understanding functions and relations. Instead of trying to work on a concept that is complex, make sure you have a good understanding of the basic concepts it draws from.

Chapter 6: Exponential and Logarithmic Functions

Exponential Functions

Exponential functions are functions that have the form $f(x) = b^x$, where b is the base and x is an exponent.

The graphs of exponential functions give us an interesting shape. In order to show and explain this, let's construct an exponential graph of our own.

Let's say $f(x) = 2^x$. For this example, I'll plug in -5, -3, 0, and 2 for x. In order to construct a table for this graph, we may have to review a basic concept: negative exponents.

What do you do when you plug in -5 for x and do 2^{-5}?

When dealing with negative exponents, remember:

$b^{-x} = 1/b^x$

So

$2^{-5} = 1/(2^5) = 1/(2 \cdot 2 \cdot 2 \cdot 2 \cdot 2) = 1/32$

For practice, let's fill in the next value in our table: x = -3

If x = -3, $f(-3) = 2^{-3}$

We now know that

$2^{-3} = 1/(2^3)$

So

$2^{-3} = 1/(2 \cdot 2 \cdot 2) = 1/8$

It is important to remember that any number to the power of zero is one.

$b^0 = 1$

Now we can construct a table.

x	f(x)
-5	1/32
-3	1/8
0	1
2	4

If we graph this information, we get:

When we graph $f(x) = 2^x$, we get a line that seems to be very close to y = 0 on one end. As the graph moves away from the point very close to y = 0, it seems to shoot up faster and faster.

If we traveled down the graph as it approached y = 0, it would continue to curve towards zero but would never get there. This is because negative values of x will make $f(x)$ very, very small, but $f(x)$ will never equal zero or a negative number. The line that $f(x) = 2x$ moves towards but never reaches is called an asymptote. Asymptotes are lines that graphs approach but never reach. As evidence of the fact that $f(x) = 2x$ will never reach y = 0, just consider what $f(x)$ is at x = -50. (At x = -50, $f(x)$ is 1/(1.12589991 × 1015). This is very small, but it is bigger than zero.)

On the other end of the graph, we see the opposite pattern; the graph seems to skyrocket away from y = 0. This rapid upwards movement is called exponential growth. As you could probably guess, the upward movement of this function gets more dramatic as x increases.

For example, when x = 2, $f(x) = 4$. When x = 3, $f(x) = 8$. In this case, increasing x by one resulted in an increase of 4 in $f(x)$. Let's see what happens when we increase by one number again, this time going from 3 to 4. As we have said, when x = 3, $f(x) = 8$. When x = 4, $f(x) = 16$. This time a difference of 1 in x yields an increase of 8.

For practice, try to graph the exponential function $f(x) = 1.5^x$.

Solution:

To graph $f(x) = 1.5^x$, start by creating a table of values.

To capture the curve of the graph, let's choose a good range of values.

x = -4, -2, 0, 2, and 4 are good.

When we plug in these values, we get this table of values:

x	f(x)
-4	16/81
-2	4/9
0	1
2	2.25
4	5.0625

To get $f(x)$ when x = -4, remember that $1.5^{-4} = 1/(1.5^4)$

As you can see from the table, $1.5^4 = 5.0625$.

This means that $1.5^{-4} = 1/5.0625 = 16/81$.

You get 16/81 by reducing 1/5.0625.

To get $f(x)$ when x = 0, remember that any number to the power of 0 = 1.

If you graph this information, you will get a graph that looks like this:

Exponential functions can also express something called exponential decay. In order to do this, though, we need to introduce a new formulation of the exponential function. Remember, the one we started with was $f(x) = b^x$.

This new formulation is $f(x) = ab^x$.

In this formulation, a and b are constants. If we create a function using this formulation of the exponential function, and b is greater than zero but less than one, we will get a graph that demonstrates exponential decay.

To show this, we will create a graph of $f(x) = (2)(.5^x)$

To do this, we will need to create a table. For x, we will use -2, -1, 0, 1, and 2.

If we plug in these values, we get this table:

x	f(x)
-2	8
-1	4
0	2
1	1
2	.5

Graph it and you get this:

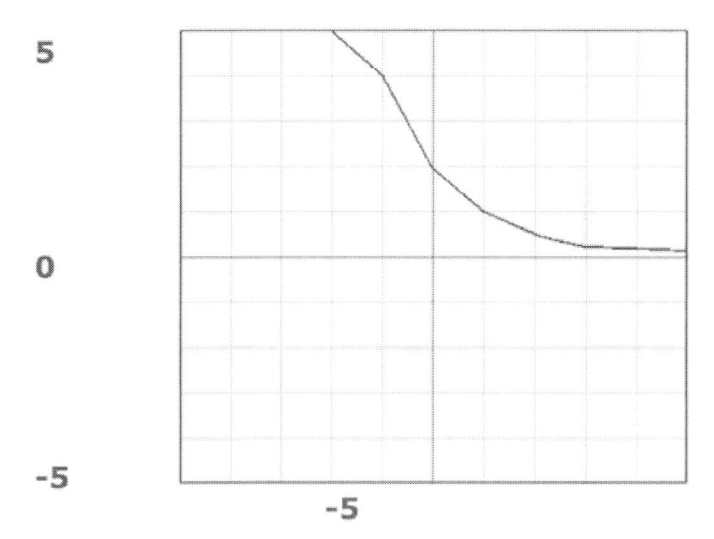

As you can see, the line now curves towards y = 0 as x gets bigger. When we see a graph like this, where $f(x)$ gets much smaller as x gets bigger, we say that it decays exponentially.

We got a graph that demonstrated exponential decay by choosing a function that was formulated such that $f(x) = ab^x$ and b was bigger than 0 but less than 1. Had we chosen to make a more than zero and b more than 1, we would have a graph that demonstrated exponential growth.

Practice:

1. $5^{-3} =$
2. $5^0 =$
3. Without constructing a graph, would $f(x) = 5^x$ demonstrate exponential decay or growth?
4. Without constructing a graph, would $f(x) = (5)(.5^x)$ demonstrate exponential decay or growth?
5. Without constructing a graph, would $f(x) = (2)(5^x)$ demonstrate exponential decay or growth?

Solutions and explanations:

1. Remember: $b^{-x} = 1/b^x$

    That means that $5^{-3} = 1/5^3 = 1/125$

2. Remember: any number to the power of $0 = 1$

    That means that $5^0 = 1$.

3. This would demonstrate exponential growth. If we think of this problem in terms of the a function in the form $f(x) = ab^x$, we could rewrite $f(x) = 5^x$ as $f(x) = (1)(5^x)$. As you can now see, a is greater than 0 and b is greater than 1, so this will demonstrate exponential growth.

4. Exponential decay, because b is greater than 0 and less than 1.

5. Exponential growth, because a is greater than 0 and b is greater than 1.

Logarithmic Functions

To understand logarithmic functions, we must first grasp the concept of logarithms.

The basic principle behind logarithms is very simple. A logarithm is an exponent written differently.

The formal way of saying this is:

If

$b^x = a$

then

$\log_b a = x$

The first statement, $b^x = a$, is written as an exponential equation.

The second statement, $\log_b a = x$ is written as a logarithmic equation.

Notice that the only real difference between the exponential equation and the logarithmic equation is the placement of the numbers.

Let's put some numbers in here to figure out how logarithms work.

Starting with something familiar, let's take $b^x = a$ and plug in some numbers that make sense and are easy to work with.

$3^2 = 9$, right?

So $b = 3$, $x = 2$, and $a = 9$.

Plug that into $\log_b a = x$, and we get $\log_3 9 = 2$.

Now, using this simple pattern, let's try to work backwards.

Try to find $\log_2 4 =$

If the answer isn't obvious, that's okay. We're going to approach this problem two ways. So if the first explanation doesn't help, don't worry! Another explanation is on the way.

Okay, so we saw that $\log_3 9 = 2$, right? Well, let's imagine that we didn't have all the variables filled out and that we only had $\log_3 9 =$

One way to think about this is to ask, "To what power do I need to raise 3 to get 9?" The answer is 2, because $3^2 = 9$.

So when we see $\log_2 4 =$ we can ask, "To what power do I need to raise 2 to get 4?" The answer is 2.

If you are still having trouble with this, here is another way to think about it:

Take the original problem, in this case $\log_2 4 =$, and label the parts. $2 = b$, $4 = a$, and x is unknown (so we can just leave it as x.) Now plug this into exponential equation ($b^x = a$). When we do this, we get:

$2^x = 4$

$x = 2$ because $2^2 = 4$.

That's why logarithms are just exponents expressed in a different way.

Still, logarithms can take some getting used to. So let's take a moment to do some exercises that will help us remember this.

Take a look at this table:

Exponential	Logarithmic
$2^3 = 8$	$\log_2 8 = 3$
$5^2 = 25$	$\log_5 25 = 2$
$7^3 = 343$	$\log_7 343 = 3$
$8^2 = 64$	$\log_8 64 = 2$
$12^2 = 144$	$\log_{12} 144 = 2$

Complete your own table.

Exponential	Logarithmic
$3^2 = 9$	
$2^5 = 32$	
$3^7 = 2,187$	
$2^8 = 256$	
$b^x = a$	

Answers:

Exponential	Logarithmic
$3^2 = 9$	$\log_3 9 = 2$
$2^5 = 32$	$\log_2 32 = 5$
$3^7 = 2,187$	$\log_3 2,187 = 7$
$2^8 = 256$	$\log_2 256 = 8$
$b^x = a$	$\log_b a = x$

Logarithmic functions are functions of the form $f(x) = \log_b x$.

We saw that logarithms are exponents expressed differently, so we might expect the graph of a logarithmic function to have a shape related to that of an exponential function.

That's right. The logarithmic function $f(x) = \log_b x$ is actually the inverse of its corresponding exponential function, $f(x) = b^x$.

To demonstrate this, let's construct a table for $f(x) = \log_2 x$. After we construct this table, we can compare it to the table of $f(x) = 2^x$ that we constructed in the previous section.

Recall that in our table for $f(x) = 2^x$, we got:

x	f(x)
-5	1/32
-3	1/8
0	1
2	4

Now, look what happens when we create a table for $f(x) = \log_2 x$

x	f(x)
1/32	-5
1/8	-3
1	0
4	2

That was easy, right? Since $f(x) = 2^x$ and $f(x) = \log_2 x$ are inverse functions, the two columns just switched. If you do not understand how we got those values, hold on just a second. We'll cover that shortly. For now, direct your attention to the graph of $f(x) = \log_2 x$:

Notice that the graph appears to be a reflection of $f(x) = 2^x$:

Now, if you didn't realize that $f(x) = \log_2 x$ was the inverse of $f(x) = 2^x$, and that you could just stitch the x and $f(x)$ columns, that's okay! You're going to learn how to find every value in the table for $f(x) = 2^x$.

Let's start with x = 1/32 and work our way down.

If x = 1/32, $f(1/32) = \log_2 1/32$

Now, you can ask yourself, "To what power do I need to raise 2 to get 1/32?"

Let's break it down. 1/32 is a fraction, so it looks like we're dealing with a negative exponent. Now ask, "To what power do I need to raise 2 to get 32?"

$2^1 = 2$

$2^2 = 4$

$2^3 = 8$

$2^4 = 16$

$2^5 = 32$ – that's it!

To answer the initial question, we must raise 2 to the power of -5 to get 1/32.

Now that we have this under our belt, figuring out $f(x)$ when x = 1/8 should be quite a bit easier. Since $2^3 = 8$, we can surmise that $2^{-3} = 1/8$. Therefore, $f(x) = -3$ when x is 1/8.

How about when x = 1? Well, if we plug this into $f(x) = \log_2 x$, we get $f(1) = \log_2 1$. To what power must we raise 2 to get 1? Zero, because any number raised to the power of 0 is 1.

Finally, how do we solve $f(x) = \log_2 x$ when x = 4? If you have solved or understood any of the preceding problems, this one shouldn't be too hard. To what power must we raise 2 to get 4? 2! So $f(x) = 2$ when x = 4.

Conclusion

There you have it! Together we've covered the fundamentals of algebra, covering everything from algebraic expressions to solving systems of inequalities. By now, you should be an expert on all things algebra. If not, there's no reason to be alarmed. Algebra can be difficult, especially when you're getting started. However, you are capable of doing it and even mastering it. If you find yourself feeling unsure of this, take a step back and think about what you don't understand. Start back on the most basic level of the topic you're struggling with and work your way up. This guide has stressed the connection between the different topics covered, so it should be easy to start with the fundamentals and build your way up.

Good luck!

Printed in Great Britain
by Amazon.co.uk, Ltd.,
Marston Gate.